Integer Island

Thomas K. Johnson

DEDICATION

I dedicate this book to my Aunt Jessie L. Neat. A woman of faith who will be dearly missed by all of her family!

CONTENTS

ACKNOWLEDGMENTS

I would like to thank my seventh grade math teacher Ms. Perkins for showing me so much enthusiasm when I first wrote this story in her math class. You made me believe my work was really good when I thought it was just ok. I would also like to acknowledge my seventh and eighth grade home room teacher Ms. Forbes for reading us books that caught our interest and for making it enjoyable. I would also like to acknowledge my fifth through eighth grade computer teacher Ms. Kula for always encouraging me to write and letting me type and print my stories in the computer room. I want to finally acknowledge my aunt Jessie for all of the books she used to buy and read to me. She truly helped to inspire imagination within me.

The story of the Times Stone takes place in Curriculum Country which is divided into four regions: Tint Town, Symbol City, Vector Village, and Integer Island. All of the colors lived in Tint Town, the letters lived in Symbol City, the shapes lived in Vector Village and, finally, the numbers resided in Integer Island. The numbers lived split between Odd numbers one through ninety-nine and Even numbers two through one hundred. Centuries ago, a Great War known as the Equation War erupted on Integer Island because the Odds wanted to take over all of Curriculum Country. However, before the Odds could try, the Evens stopped them by using the

Sign Stones. There were four Sign Stones and each stone was located in a different Sign Temple. The Minus Stone was located in the Subtraction Temple, the Plus Stone was in the Addition Temple, the Divide Stone was in the Division Temple and the Times Stone was in the Multiplication Temple. When two numbers placed their hands on a Sign Stone, they performed the action of the stone they touched. Thus, the numbers would either add, subtract, divide or multiply with each other. After the war was over, the Minus, Plus, Divide and Times Stones were all believed to be destroyed. Many numbers were deleted and only a few numbers survived the war. The only numbers who survived the war were numbers three, five, seven, and nine from the Odds; two, four, six, eight, and ten from the Evens, and Zero; who was neither an odd nor an even number. Because Zero had no numerical value, all of the other numbers showed Zero no respect and viewed him as weak and inferior. Zero was always trying to find his place within Integer Island. He deeply loved being a digit but felt unaccepted by both the Odds and Evens. After the Great Equation War, Zero lived amongst Odds with numbers three, five, seven, and nine on their side of the island. The Odds were very selfish numbers and always looked for an opportunity to take advantage of a lesser valued integer. When Zero first moved to the Odds' half of the island, he built himself a small hut in which to live. After a few decades of living there, number three came to Zero and told him he couldn't live there anymore because he needed a place to store all of his equal signs and Zero's hut was the perfect size. Zero tried to fight Three off, but wasn't strong enough to beat him. Three laughed uncontrollably as he watched Zero walk away homeless.

Zero was disheartened, but stayed on the Odds' half of the island and decided to build another hut. After a few years of living there, number five came over to the hut and demanded that Zero give him his hut. Zero replied, "No, why do you want my hut?" Five told Zero he needed a new hut to store his pile of parenthesis and that his hut was a perfect size. Zero told Five to leave, but Five grabbed Zero, threw him out of the hut, and then said "Thank you for understanding," as he laughed at Zero's misfortune. Later that night as Zero wandered around the island trying to figure out his next move, he noticed all of the Odds going over to Nine's house. Zero quietly snuck over to the house and crouched underneath the window to hear what was happening. As he listened, he heard Nine telling the Odds to keep on the look–out for the Times Stone which was buried deep beneath the Multiplication Temple. "As soon as we find the stone we're going to ambush the Evens," Nine gloated. Zero was surprised because the Odds and

Evens swore a truce between each other after the war and agreed to start a new era on Integer Island towards peace and unity. The numbers also agreed to work together to repair the broken temples. The Odds did not want peace and unity; rather, they wanted power and to rule over not just Integer Island but all of Curriculum Country. Nine told the other Odds that not all of the Sign Stones were destroyed long ago during the war. He told the Odds that number eleven had buried the Times Stone deep beneath the Multiplication Temple because the Odds were losing the war. Eleven decided to bury the stone just in case they lost the war completely so that in the future, the Odds could recover the stone and take revenge against the Evens by surprise. Nine continued explaining what happened with Eleven, "Right before Twenty-Two from the Evens deleted Eleven, number eleven was able to inform me as to where the stone was buried. Twenty-Two knew

Eleven had the stone and demanded he give it to him but Eleven lied and said he destroyed the stone. After that, Twenty-Two deleted him. Twenty-Two was later reduced to an odd when Twenty-One used the Subtraction Stone and reduced him to one; who was deleted by four". After Nine told the Odds the history, he said, "After all these years of rebuilding the temple, we have been getting closer to the Times Stone. Be ready, our time is coming soon." Upon hearing of what the Odds were planning to do, Zero made his way to the other side of the Island where the even numbers two, four, six, eight, and ten stayed.

The Evens were power hungry and respected strength. They constantly tried to figure out ways to increase their value without the use of the Addition Stone or the Times Stone. In fact, the Evens practiced a technique called the times dance that allowed them to multiply with each other without the use of a

Times Stone. Nevertheless, the Evens had no desire to rule over anyone else and wished for the Odds and Evens to live peacefully together in unity. To them, your numerical value is everything as it defines who you are. Zero made his way to Ten's house and banged on the door. When Ten opened the door, he looked at Zero only to quickly look away, rolling his eyes stating, "Oh it's you – the no count. What do you want?" Zero explained that he had no place to live because the Odds kept kicking him out of his huts. In his sorrow, he began to put himself down even more by saying, "Other numbers treat me badly because I have no numerical value: if you add me with another number the number stays the same, if you subtract me, the number stays the same, and if you divide me by another number it's undefined. The only power I have is if – " "HA HA HA" ten interrupted and laughed uncontrollably. "You have NO power. I don't mind if you live amongst us, just don't get in our way… wait, you can't – you're powerless HA HA HA!" Zero walked away frustrated at Ten's mocking response towards him, but was encouraged that he now had a stable place to live. Zero built a hut and lived peacefully, but soon became lonely. Although he wasn't being kicked out of his hut like he did when he lived with the Odds, the Evens treated Zero as if he didn't exist. The only number who acknowledged Zero's presence was Two, but only because they worked together to rebuild the Subtraction Temple. Besides this interaction, Zero was completely ignored. Since Zero had so much time to think, he began to remember the Odds' plan to find the Times Stone and overthrow the Evens. He wasn't sure if he should warn the Evens of the Odds' treachery because both sides have treated him with disrespect. Zero thought to himself, "I could easily vanish and no one would care. Why should I try and help the Evens? What's in it for me?". Then Zero let out a huge sigh and said to himself, "It's not in my best interest, but it is the right thing to do". Zero left his hut and went to Tens house. All of the Evens were outside practicing their times dance because they were obsessed with trying to figure out how to increase their value without the use of the Multiplication Stone or the Addition Stone but could never achieve it. Zero ran to Ten and tried to explain what the Odds had planned for them, but Ten ignored Zero and continued with his routine. Zero then ran to Eight and explained the same thing, but Eight barely glanced at Zero. He then tried to explain it to Six, who looked at Zero, rolled his eyes, and continued with his work. Zero went over to Four to talk to him but Four gave Zero a very angry look and motioned for him to leave him alone. Finally, Zero tried to reason with Two, who did listen to everything Zero had to tell him. Two replied, "The Odds wouldn't dare test us like that. Ten has more value than nine! They were probably just trying to make a fool out of you." After that, Zero gave up on trying to convince the Evens and walked far away from Integer Island. For days, Zero was missing and no one even noticed he was gone, except for Two because he was short a partner for rebuilding the Subtraction Temple. Every day, the numbers would split up into teams to rebuild the Sign Temples that were destroyed during the war. Numbers Zero and Two worked on the Subtraction Temple, Three and Four worked on the Addition Temple, Five and Six worked on the Division Temple, while Seven, Eight, Nine, and Ten worked on the Multiplication Temple. Since the Evens desired peace throughout Integer Island and had no desire to subjugate the Odds, they actually didn't care what the Odds did as long as they didn't try to overthrow

Curriculum Country. The Evens always wanted to have one even who was bigger than an odd as they worked on the temple together to remind the Odds that the Evens are more powerful than they are.

The next day, the Odds and Evens gathered together for their daily work as they usually did. As number seven searched through the rubble of the Multiplication Temple, he found the Times Stone. He immediately picked it up and yelled to number nine, "I found the stone, now we can put our plan into action!" Number eight was standing nearby and immediately overpowered Seven and took the stone. Just then, Nine came and wrestled the stone away from Eight as Ten was too far away to get to Nine in time. At that moment, Nine and Seven placed their hands on the stone and became sixty-three! Sixty-Three began to thrash Eight and Ten then walked over to Three. Sixty-Three and Three touched the stone and became one hundred and eighty nine and yelled, "I am invincible!" Then One Eighty Nine walked over to Six and deleted him. "NO!", Ten yelled, as he was helpless to stop him. While One Eighty Nine began to laugh hysterically, number four quietly snatched the Times Stone from One Eighty Nine and threw it to Ten. Hastily, Ten and Eight touched the stone and became eighty, but the Evens still weren't strong enough to defeat one hundred and eighty nine. Then Eighty and Four touched the stone and became three hundred and twenty! Three Twenty and One Eighty Nine began to battle but One Eighty Nine was not able to beat Three Twenty and was easily overwhelmed. Out of frustration for not being able to defeat Three Twenty, One Eighty Nine glanced at number two and then barked to Three Twenty, "I can't beat you – but I can crush him!" Three Twenty rushed as quickly as he could to stop One Eighty Nine, but was too late.

One Eighty Nine got to Two first and deleted him. Three Twenty was so angry that he dropped the stone in order to thrash One Eighty Nine. He beat him so badly that One Eighty Nine was moments away from deletion. Just then, number five picked up the stone and put it in One Eighty Nine's hand. Five and One Eighty Nine then became nine hundred and forty five! Nine Forty Five threw Three Twenty off of him and began to take his revenge.

"The Odds will finally get even, now that we are the most powerful group on Integer Island."

"The first thing I'll do is imprison you Evens and make you feel what it is like to be the weaker digit."

Nine Forty Five also destroyed the Times Stone so that no one could use it against him. He planned to torture Three Twenty for a few days before completely deleting him and then planned to wage war against the rest of Curriculum Country. Nine Forty Five built a huge castle on the side of the Evens island, and built a prison for Three Twenty to stay. Three times a day, Nine Forty Five would let Three Twenty out of the prison to try and escape, and would then tell him, "If you can beat me, I'll let you leave." However, Three Twenty could never beat Nine Forty Five, and every time Nine Forty Five would beat Three Twenty he would stand on top of him and yell in his face, "You are weak and will never be able to defeat me!" One day, Nine Forty Five decided he was done playing with the Evens and let Three Twenty out of the cell for the last time. After he beat Three Twenty this time, he was going to delete him. Three Twenty knew that Nine Forty Five was going to delete him if he lost this time, so he ran to him with all of his strength and tackled him to the ground. Nine Forty Five picked up Three Twenty with one hand and flung him into a tree. He then grabbed Three Twenty by his legs, swung him around, and threw him high into the sky. When Three Twenty was falling back down to the ground, Nine Forty Five grabbed him and body slammed him so hard that the ground split in half. Three Twenty was beaten both physically and mentally as Nine Forty Five laughed and said, "I have finally broken the Evens' pride!" As he was about to deliver the final blow and delete Three Twenty, he heard a loud voice shout, "STOP!!!" There was someone standing on top of the hill but Nine Forty Five couldn't see who it was because the sun was in his eyes. "Who dares to challenge the unbeatable Nine Forty Five?!" Just then, the person on top of the hill slid down and revealed himself. . . It was ZERO! "NO!" Yelled Three Twenty as he lay helpless on the ground, "RUN SAVE YOURSELF – YOU HAVE NO VALUE. YOU CAN'T DO ANYTHING, RUN"! Nine Forty Five put his foot on Three Twenty's chest, "QUIET!" he yelled.

"I'll get back to you after I deal with this no count." Zero looked at Nine Forty Five, "I may be a no count, but I will be a no count with courage who stands up to digits like you." Nine Forty Five walked over to Zero – who appeared confident – grabbed him and asked, "Any last words before I delete you?" Zero replied, "Just because I'm a no count doesn't mean I have no value," then he began to do the times dance. Three Twenty yelled at Zero, "What are you doing, we have never been able achieve that power!" Nine Forty Five also yelled, "What are you doing?" Zero finished the dance and then said "This" and touched Nine Forty Five. Suddenly, Zero and Nine Forty Five began to glow. Nine Forty Five began to panic, and yelled, "NOO WHAT IS HAPPENING?" Zero smirked, "I don't need the Times Stone to multiply. Don't you know anything multiplied by Zero is ZERO!" At that very moment, Zero and Nine Forty Five merged and became Zero. Three Twenty was amazed and mumbled, "HOW, How did you. . . I don't understand." Zero explained, "I used to watch the Evens practice the times dance and then I started practicing it by myself. One day I achieved its power but I didn't bother to tell you because I thought you wouldn't listen." Three Twenty began to cry and said "Thank you for saving me from deletion, Zero. I will never judge a digit by its numerical value ever again, but by its character."

ABOUT THE AUTHOR

Thomas Kareem Johnson was born and raised in the South Bronx and has also lived in Manhattan's Lower East Side. He has dedicated his experience to serving at risk youth and autistic children for the past eight years as a Teacher's Aide and Afterschool Tutor and Mentor. As a strong advocate for youth literacy, he believes that reading comprehension is a key factor in self –reliance and enjoys exploring ways to elevate children's love of learning. Kareem has published work in the "Anthology of Poetry by Young Americans" and has also published his first Children's book, Oogly Glook. Integer Island is his second Children's book.